Printed in Poland
by Amazon Fulfillment
Poland Sp. z o.o., Wrocław

24917194R00069